SCIENTIFIC AMERICAN EDUCATIONAL PUBLISHING

SCIENTIFIC AMERICAN INVESTIGATES LIFE CYCLES

THE STRANGE LIFE CYCLE OF AN OCTOPUS

ANNA MCDOUGAL

Published in 2025 by The Rosen Publishing Group
in association with Scientific American Educational Publishing
2544 Clinton Street, Buffalo NY 14224

Library of Congress Cataloging-in-Publication Data
Names: McDougal, Anna, author.
Title: The strange life cycle of an octopus / Anna McDougal.
Description: Buffalo, NY : Scientific American Educational Publishing, [2025] | Series: Scientific American investigates life cycles | Includes bibliographical references and index.
Identifiers: LCCN 2024011231 (print) | LCCN 2024011232 (ebook) | ISBN 9781725350946 (library binding) | ISBN 9781725350939 (paperback) | ISBN 9781725350953 (ebook)
Subjects: LCSH: Octopuses–Juvenile literature. | Octopuses–Life cycles–Juvenile literature. | Life cycles (Biology)–Juvenile literature.
Classification: LCC QL430.3.O2 M43 2025 (print) | LCC QL430.3.O2 (ebook) | DDC 594/.56–dc23/eng/20240320
LC record available at https://lccn.loc.gov/2024011231
LC ebook record available at https://lccn.loc.gov/2024011232

Portions of this work were originally authored by Therese Shea and published as *The Bizarre Life Cycle of an Octopus.* All new material in this edition authored by Anna McDougal.

Designer: Tanya Dellaccio Keeney
Editor: Caitie McAneney

Photo credits: Cover K A Studio/Shutterstock.com; p. 5 Olga Visavi/Shutterstock.com; p. 7 Divelvanov/Shutterstock.com; p 9 kkshxt/Shutterstock.com;p. 11 Andre-Johnson/Shutterstock.com; p. 13 Alex Navarro Menal/Shutterstock.com; p. 15 Konstantin Novikov/Shutterstock.com; p. 17 pr2is/Shutterstock.com; p. 19 Henner Damke/Shutterstock.com; p. 19 Agarianna76/Shutterstock.com.

Printed in the United States of America

CPSIA compliance information: Batch #CSSA25. For Further Information contact Rosen Publishing at 1-800-237-9932.

CONTENTS

Words in the glossary appear in **bold** type the first time they are used in the text.

INCREDIBLE INVERTEBRATE

Octopuses are some of the most incredible animals in the ocean. You can identify them by their soft body, or mantle, and eight long arms. The octopus is a kind of **mollusk** called a cephalopod. "Cephalopod" means "head-foot." An octopus's "feet" (or arms) attach to its "head" (or mantle).

Octopuses are very smart. They can also fit into tight spaces. That's because they are invertebrates, or animals without a backbone. Their fun features aren't the only cool things about them, though. Octopuses have incredible life cycles too.

Octopuses breathe with gills, just like fish. Water flows into the mantle through the gills and out through a tube called a siphon.

FUN FACT

OCTOPUSES HAVE NO BONES AT ALL! THE ONLY HARD PARTS ON THEIR BODIES ARE THEIR BEAKS.

THE CYCLE STARTS

An octopus's life cycle starts when a female and male **mate**. The male octopus has one arm that looks a bit different from the other seven. He uses this arm to give the female special cells to **fertilize** her eggs. Male octopuses die soon after mating.

Scientists think that the female octopus makes a special **chemical** in her body that interests a mate—and keeps him from eating her.

This female octopus keeps her eggs close.

FUN FACT

THE BLANKET OCTOPUS HAS A STRANGE MATING HABIT. THE MALE TAKES OFF HIS ARM AND GIVES IT TO THE FEMALE!

A MOTHER AND HER EGGS

A female octopus lays tens of thousands of eggs at once. Then she weaves them all together with a string. She hangs these strings of eggs from the roof of her den.

The mother octopus guards her eggs for weeks or months. She blows water at them so they get **oxygen**. She keeps them clean using the **suckers** on her arms. She spends all her time with the eggs and doesn't eat. Her job is done when her eggs break open. Then, she dies.

Caring for her eggs causes a mother octopus to slowly weaken.

FUN FACT

SOMETIMES A MOTHER OCTOPUS STAYS WITH HER EGGS FOR UP TO A YEAR!

OCTOPUS BABIES

When octopuses are born, they're called paralarvae. They swim up and join the **plankton** drifting on the ocean's surface. They eat anything they can catch. Scientists don't know much about the lives of young octopuses. They do know that very few live to become adults. While baby octopuses float, they're in danger of being eaten.

After a few weeks, most baby octopuses, now called juveniles, swim to the ocean floor. They may make a den out of rocks, which is called a garden.

Baby octopuses are food for whales and fish when they float.

FUN FACT

PREDATORS OF JUVENILE OCTOPUSES INCLUDE SEALS, SHARKS, EELS, DOLPHINS—AND EVEN OTHER OCTOPUSES!

AWESOME ADAPTATIONS

Octopuses have amazing **adaptations** to stay alive. Some kinds of octopuses change color and shape to match their surroundings. They have special skin cells and **muscles** that work quickly to blend in with rocks and other things in the ocean. That's called camouflage.

When an octopus fears a predator, it may release a dark inky liquid from its body. An ink cloud can hide the octopus and help it escape. It may also hurt the eyes of the predator and throw off its senses of smell and taste.

Can you spot the octopus? It has taken on the color and texture, or feel, of its surroundings.

FUN FACT

IF A PREDATOR CATCHES AN OCTOPUS'S ARM, THE OCTOPUS CAN LEAVE THE ARM BEHIND! THE OCTOPUS CAN THEN GROW ANOTHER ARM.

ON THE HUNT

Octopuses are great hunters. They use their super senses and features to help them catch a meal. Octopuses have excellent eyesight and mostly hunt at night. They use their long arms to reach creatures in tight spaces. They may use their ink cloud to hide before attacking an animal.

Octopuses have a sharp beak on their underside to bite animals and break through shells. **Venom** in their spit breaks down the insides of an animal so it's easier to eat. The blue-ringed octopus has venom so deadly that it can kill people!

FUN FACT

OCTOPUSES LIKE TO EAT CRABS, LOBSTER, CLAMS, AND SHRIMP. A GIANT PACIFIC OCTOPUS MAY EAT BIRDS AND SHARKS TOO!

The suckers on an octopus can taste whatever they touch so the octopus knows if it has found food.

GROWING UP

About 300 species of octopus live on Earth, and they all grow to different sizes. Most octopuses grow to be about 1.6 feet (50 cm) long. They usually weigh less than 22 pounds (10 kg). However, there are octopuses much smaller and much bigger.

The giant Pacific octopus can grow to up to 30 feet (9 m) long and weigh hundreds of pounds. The smallest octopus is the California octopus, which is only about 1 inch (2.5 cm) long.

Female giant Pacific octopuses have 280 suckers on each arm!

FUN FACT

ADULT OCTOPUSES ARE SOLITARY, WHICH MEANS THEY LIVE ALONE. THEY ARE ALSO TERRITORIAL, WHICH MEANS THEY LET NO ONE IN THEIR SPACE.

THE BEGINNING OF THE END

Most of an octopus's life is spent hunting and hiding. After they've grown, they look for a mate. That means the beginning of the end of the life cycle of the octopus.

The last stage of the cycle is called senescence (sih-NEH-suhns). It happens to males after mating and to females after laying their eggs. They stop eating, are unaware of predators, and don't care for themselves in other ways too. Soon after, they die.

Octopuses usually live for only about a year.

FUN FACT

THE GIANT PACIFIC OCTOPUS LIVES THE LONGEST OF ALL SPECIES. THEY LIVE BETWEEN THREE AND FIVE YEARS.

BIG BRAINS

Scientists have long studied these interesting invertebrates. They've found that octopuses are the smartest invertebrates on Earth. Their huge brains have the ability to remember and learn things. They take things apart and solve problems. Some can open jars.

You can go to an aquarium to check out an octopus up close. Aquariums have to take special care with their octopuses—they're known to escape! These smart, special animals are always up to something new, from hiding in a bottle to solving a maze.

LIFE CYCLE OF AN OCTOPUS

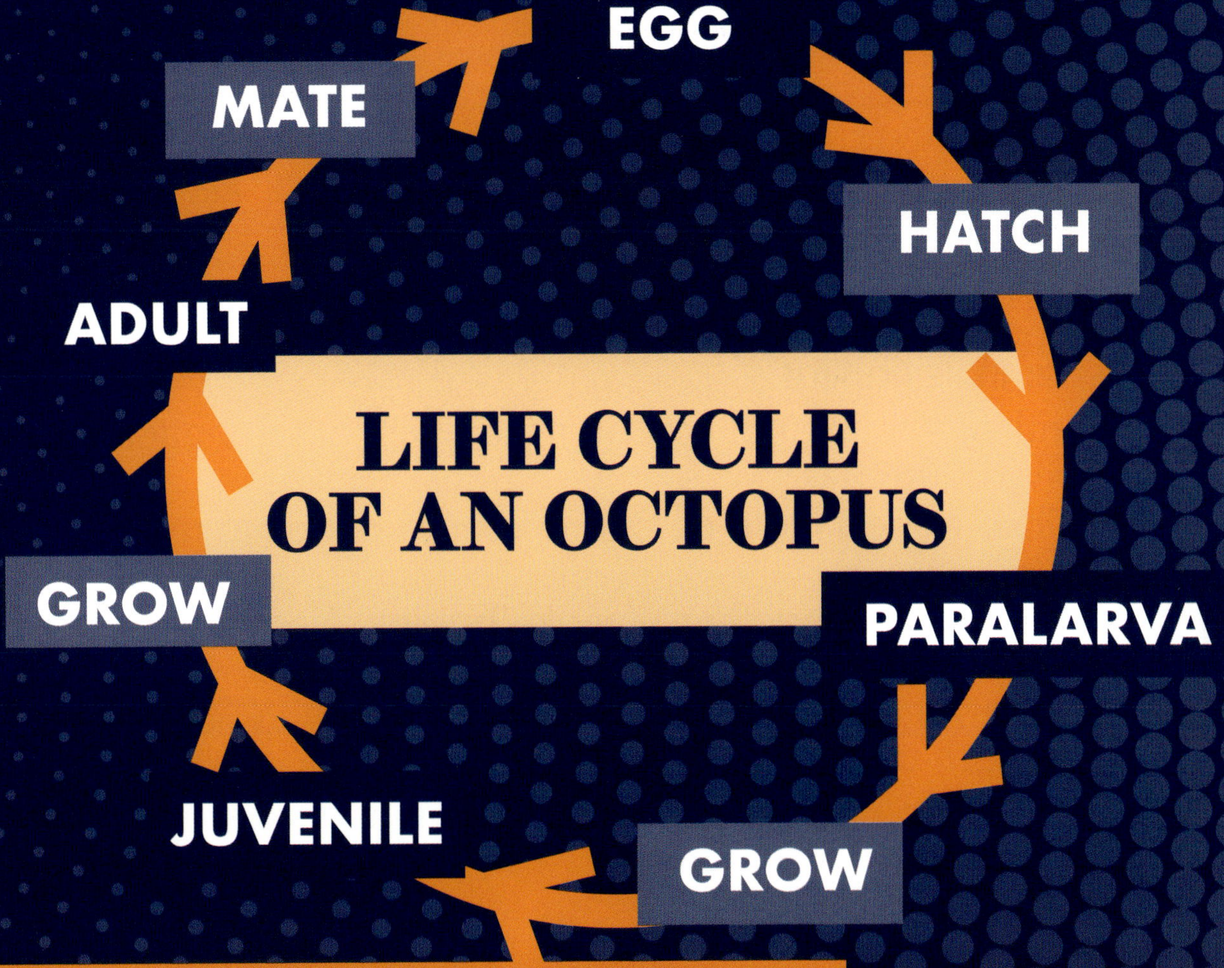

Octopuses can hide in anything that's bigger than their beak since that is the only hard part on them.

GLOSSARY

adaptation: A change in a living thing that helps it live better in its habitat.

chemical: Matter that can be mixed with other matter to cause changes.

fertilize: To add male cells to a female's eggs to make babies.

mate: To come together to make babies. Also, one of two animals that come together to make babies.

mollusk: An animal that lacks a backbone and has a soft body, such as a snail, clam, or octopus.

muscle: One of the parts of the body that allows movement.

oxygen: A gas that is necessary for people and animals to breathe.

plankton: A tiny plant or animal that floats in the ocean.

predator: An animal that hunts other animals for food.

sucker: Part of an animal used for holding on to things.

venom: Poisonous matter created by an animal and passed on by a bite or sting.